TREE, SHRUB, AND VINE SEEDS
A Pictorial Field Guide

I0413997

Terry A. Woodger

Universal-Publishers
Boca Raton

Tree, Shrub, and Vine Seeds:
A Pictorial Field Guide

Universal-Publishers
Boca Raton, Florida
USA • 2011

ISBN-10: 1-61233-044-4
ISBN-13: 978-1-61233-044-0

www.universal-publishers.com

Library of Congress Cataloging-in-Publication Data

Woodger, Terry A.
    Tree, shrub, and vine seeds : a pictorial field guide / Terry A. Woodger.
    p. cm.
Includes bibliographical references and index.
ISBN-13: 978-1-61233-044-0 (pbk. : alk. paper)
ISBN-10: 1-61233-044-4 (pbk. : alk. paper)
1. Trees--Seeds--Harvesting. 2. Shrubs--Seeds--Harvesting. 3.
Climbing plants--Seeds--Harvesting. 4. Seeds--Cleaning. 5. Trees--
Seeds--Storage. 6. Shrubs--Seeds--Storage. 7. Climbing plants--Seeds--
Storage. 8. Trees--Seeds--Pictorial works. 9. Shrubs--Seeds--Pictorial
works. 10. Climbing plants--Seeds--Pictorial works. I. Title.
SB118.3.W69 2011
631.5'21--dc23

                                                        2011032367

ACKNOWLEDGMENTS

A book of this nature can not be written without the assistance of family and friends. I would like to acknowledge the following people who assisted in many ways to help make this book a reality.

First, I'd like to thank my wife and children for their support and encouragement, without which this book would never have been completed.

I'd also like to thank the following people for their assistance:

Andrew Leighton
Barry Billington
Clive and Kathy Grimshaw
Colin and Shirly Cattle
Eddy and Marion Pettifor
Graham and Kathy Evans
Jacqueline Weight
Ron and Lynda Roundhill
Tim and David Evans

Disclaimer:
Plants have many ways in which they protect themselves from damaging organisms. This protection is found in thorns, sap, toxins, etc. Although the collection, cleaning, and storage of seed can be a rewarding experience, the author takes no responsibility for injury or illness that results from these activities.

1

CONTENTS

INTRODUCTION

Seeds are an exciting and beautiful component of a productive garden. Beautiful gardens that provide a great sense of fulfillment can be created from just a handful of these treasures.

Although some of these plants can propagate by other means, such as bulbs or division, seeds are the principal way in which plants reproduce.

This book covers the basics involved in the collection, cleaning and storage of seeds. Although bulbs, corms, and other plant parts can be collected and stored, they are not covered here, so as to not detract from the focus of this volume.

As the number of plants grown in gardens is truly staggering, it is impossible to cover them all. In this field guide, we discuss the most common plant families, including examples of the types of seeds that may be encountered. Where possible, several genera within each family are discussed.

This book uses a system whereby plant family names are all written in capitals (STRELITZIACEAE), the common names that are not written within the text are in bold (**Bird of Paradise**), and the botanical names are written in italics (*Strelitzia nicolai*).

In botany, it is the characteristics of the flowers that determine the genera and family to which a plant belongs. This can become extremely complicated, so this field guide makes no mention of the flower types or their individual differences.

Also discussed are a number of methods that can be utilized in the collection of seeds. No one method can be used to collect them all, so different techniques have been developed over time to successfully gather all of the species that are encountered, both in the home garden as well as in the field.

The same development of techniques applies to the cleaning of seeds. There are a number of ways in which common household items can be used effectively to clean seeds. Several of these items are explained in Chapter 3: How to Thresh and Clean Seed.

Storing seeds for use next season can be fraught with hidden problems, such as molds and seed-borers. Chapter 4: The Storage of Seed explains some appropriate methods and procedures that should be followed to avoid disappointment and loss of seed.

Most seeds collected from the garden are suitable for storage from one year to the next, and many of these can be successfully stored at home for many years.

Some things to consider when collecting seeds are the quantity and the number of plants from which they are collected.

Collecting seed from only one fruit on one plant over several seasons can have unforeseen consequences, such as the collection of seeds from small fruit, especially from only one plant. This can produce smaller plants with smaller fruit. This is often referred to as "line-breeding" and can, over time, lead to complete genetic breakdown and a loss of viability.

The collection of seed is not always a simple matter, and it is very important that, when possible, you gather seed from several plants to maintain their genetic stability.

Size does matter, and in the case of growing flowers, it is always best to collect seed from the biggest fruit or healthiest plants.

The exception would be if you are after a specific genetic trait. For example, seed selection would vary for a plant desired as a bush if that plant is usually a climbing variety.

Now let's look at the question of: what is a fruit?

When considering plants, a fruit is any structure that produces seed, but not always something that is edible. The following pages outline the types of fruit that will be encountered in the collection of seeds from trees, shrubs and vines.

One of the first things that should be learned in the collection of seed is the differences between the various fruits that you will encounter, as this will determine how to collect and clean the seed.

The methods used for the collection of seeds do not require an intimate knowledge of fruits. Knowing the basics as outlined here will aid in the collection, cleaning, and storage of viable seed.

Berry: A berry is a fruit with seeds contained within a fleshy or dry pulp. Berries contain two or more seeds, and include the tomato and kiwi fruits.

Wild Orange
Capparis mitchelii

Capsule: A capsule is a dry seed case that opens along several seams and can be papery thin to hard and woody. Some open at maturity, releasing their seeds, while others remain closed. Capsules can contain anything from one to over one hundred seeds and are found in many plant families.

African Mahogany
Khaya senegalensis

Cone: A cone is a woody structure made up of individual segments which each release one seed. Cones are found in conifers and some other flowering plants. Examples of cones are pine and sheoak

Radiata Pine *Pinus radiata*

Drupe: A drupe is a fleshy or dry fruit containing one seed, often referred to as a "stone," hence the origin of the term "stone fruits." These include the plum, nectarine, avocado, and apricot. Some seeds can produce two plants, and rarely three.

Cassowary Plum
Cerbera floribunda

Drupelet: A drupelet is a small drupe, such as an individual segment of a blackberry or raspberry.

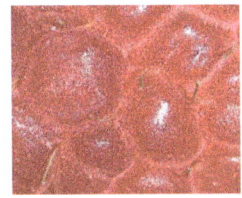

Raspberry *Rubus idaeus*

Follicle: A follicle is a dry woody capsule that opens along one side only. These may occur singularly or in clusters. The seeds can range from light, feathery seeds that can float with the wind, to thin winged papery seeds, to large beans, and anything in between.

African Tulip Tree
Spathodea campanulata

Hesperidium: A hesperidium is a berry with a thick rind and is made up of segments. These are fruits of the citrus family (RUTACEAE) and include the orange, mandarin, lemon, and many others.

Bush Lemon *Citrus* sp.

Legume: Legumes split into two equal halves upon drying. Legumes include all the beans and peas. They can be thin and papery, or thick and woody. The crab eye vine and bauhinia are examples of legumes.

Crab Eye Vine
Abrus precatramia

Nut: A nut is a woody fruit, often with one seed, that does not open readily. Most authorities lump nuts in with drupes. For the purposes of this book, they are listed separately, as nuts generally do not have a fleshy outer covering.

1cm

Hazelnut *Corylus avellane*

Pepo: The pepo is a berry with a hard rind. This includes all of the cucurbits, such as gourd, pumpkin, and cucumber.

Luffa *Luffa purgens*

Samara: A samara is a dry, winged fruit that does not open at maturity. Samaras include the maple, ash, and elm.

Rosewood *Tipuana tipu*

Stinkwood
Gyrocarpus americanus

Syconium: These fruits are from the fig family. They are round to pear-shaped, and can be shades of red or brown to black when ripe. The minute flowers are located along the inner wall of the fleshy structure. A hole in the bottom of the fruit allows entry by pollinating insects. Once pollinated, the individual flowers can grow into small single-seeded drupelets.

Fig *Ficus* sp.

CHAPTER 1
THE COLLECTION OF SEED

Seed should be collected when the weather is fine and the plants are dry.

If you have been able to collect seed pods or capsules when they are dry, then it is only a matter of cleaning. However, if the collected material is wet or damp, then mold may become an issue. Wet collected material can be spread out on a tarp in the sun to dry. If this is not possible, spread them near some sort of warmth, such as a heater, and allow them to fully dry prior to cleaning.

Removing seeds from fleshy berries and drupes is usually a simple matter if they are fully mature.

When transporting seed overseas or across borders, a declaration of the species and amounts is often required, as some species are prohibited. If the seed is not thoroughly cleaned it may be confiscated.

When seed is required for personal use and cleanliness is not so important, keep in mind that insects and molds can still become a problem, so it remains important to clean all seed as thoroughly as possible.

There are a number of ways to collect seed; it is a matter of selecting the one which works best for you and the plant species from which you are collecting. By following the methods outlined for each plant variety, the collection of viable seed should be achievable throughout the seasons.

Trees, shrubs, and vines are found throughout the world. Many are highly prized for their magnificent flowers or foliage. When utilizing these plants in landscaping, take care to plant them where they will not become a problem to building structures and underground amenities over the years to come.

Trees and shrubs are perennial plants, not annuals. They are differentiated by number of trunks, and not by their height: a tree has a single trunk, while a shrub is multi-trunked. The vines outlined in the following pages are both annuals and perennials.

Family: ANACARDIACEAE
Common name: **Cashew**
Number of genera: 70
Number of species: 600
Origin: Tropical to subtropical
Plants: Trees and shrubs

The fruit ranges from small to large fleshy drupes. The seeds can be 3 - 50mm in size, and are round, oval or kidney-shaped. They are usually white to grey in color.

Many of the fruits from this family are grown commercially in large orchards. Collect the fruits individually and carefully remove the flesh. The small seed should be washed using a sieve. However, the larger ones can be washed individually. Dry the seed thoroughly before storing. Some species cannot be stored for more than a week or so.

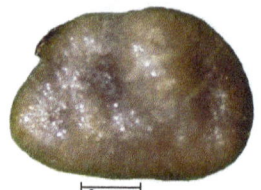

Sumac *Rhus taitansis*

Family: APOCYNACEAE
Common name: **Dogbane**
Number of genera: 220
Number of species: 2100
Origin: Tropical
Plants: Trees, shrubs, herbs, and vines

Most of these plants have a white milky sap, which is often poisonous. The fruits are fleshy berries, drupes, or pod-like capsules that are long and thin. The fruit can be harvested upon maturity or when they start to split lengthwise. The seeds can be flattened and oval to oblong, sometimes with short or long silky hairs. The seed color ranges from light brown to black. The seed is often short-lived and should be planted soon after collection.

Cassowary Plum
Cerbera floribunda

Coyol Palm
*Acrocomia
aculeata*

**Carpentaria
Palm**
*Carpentaria
acuminate*

Cheesewood
Alstonia spectabilis

Macarthur Palm
Ptychosperma macarthurii

Family: ARECACEAE, PALMAE
Common name: **Palm**
Number of genera: 200
Number of species: 3000
Origin: Tropical to temperate
Plants: Trees

The fruit can be either a dry, fibrous, or succulent drupe. The seeds can be collected as they fall from the plant or picked individually as they ripen. Alternatively, the whole bunch can be collected when most of the seeds are mature.

Clean away any fleshy material and wash the seed before spreading on a tarp to dry. Most species store well.

Blue Latan Palm
Latania loddigesii

Queen Palm
Syagrus sp.

Dwarf Stilt Palm
*Verschafffeltia
splendida*

Sugar Palm
*Arenga
caudate*

African Oil Palm
Elaeis guineensis

9

Nibung Palm
Oncosperma tigillaruh

Umbrella Tree
Schefflera cephalotes

Fiji Fan Palm
Pritchardia pacifica

Broom Palm
Thrinax parviflora

Christmas Palm
Veitchia merrillii

Family: ASCLEPIADACEAE
Common name: **Milkweed**
Number of genera: 250
Number of species: 2500
Origin: Subtropical
Plants: Perennial shrubs, herbs and vines

Many of these plants have a white milky sap which is poisonous. The fruits can be long cylindrical follicles or small ovate capsules. The seeds can be narrowly oblong, flattened, or rounded with a tuft of hairs at one end. The seeds are brown to black. The fruits are collected as they change color and begin to split. The seed detaches from the hairy tufts with no effort and little additional cleaning is required. Allow some time for the seed to dry fully before storage.

Bush Banana
Marsdenia australis

Caustic Vine
Sarcostemma viminale

Family: BARRINGTONIACEAE
Common name: **Barringtonia**
Number of genera: 20
Number of species: 450
Origin: Tropics
Plants: Trees and shrubs

The fruits are oval to squarish drupes or berries with seeds embedded within a fleshy or fibrous pulp. Collect the whole fruit once it is mature and strip away the flesh or fibers to remove the seed. Wash or clean away anything that clings to the seed.

Beach Barringtonia
Barringtonia asiatica

Family: BIGNONIACEAE
Common name: **Bignonia**
Number of genera: 100
Number of species: 800
Origin: Tropical to subtropical
Plants: Trees, shrubs and vines with some herbs

The fruits are berries or bean-like capsules with few to numerous seeds that are often winged. The seed can be collected from the capsules as the fruit splits. The seed will often remain in the capsule for a short time before falling. The fruit of the genus *Kigelia* is a berry that remains whole and is often difficult to open.

Jacaranda
Jacaranda mimosaefolia

Sausage Tree
Kigelia pinnata

Silver Trumpet Tree
Tabebuia argentia

Bower of Beauty
Pandorea jasminoides

Family: BIXACEAE
Common name: **Bixa**
Number of genera: 3
Number of species: 25
Origin: Tropical
Plants: Trees or shrubs

The fruit is a smooth or hairy capsule that splits to release numerous seeds. The seeds are kidney-shaped or partly circular. They are yellow, orange, or red, and sometimes hairy.

The seeds can be collected without difficulty from the dry capsules that often retain the seed for some time until being disturbed by the wind, passing animals, or objects.

Orellana *Bixa orellana*

Family: BOMBACACEAE
Common name: **Bombax**
Number of genera: 30
Number of species: 200
Origin: Tropical
Plants: Trees, sometimes shrubs

The fruit is usually a capsule that splits along the sides unevenly. The seeds are found along the sides of the capsule in a pithy or hairy tissue. The kidney-shaped or oval seed is often

11

smooth and hairy. Collect the whole fruit and remove the seed before sieving away the chaff.

The use of gloves is recommended, as the chaff can cause irritation to sensitive skin.

Baobab
Adansonia gregorii

Red Silk Cotton Tree
Bombax malabaricum

Kapok Tree *Ceiba pentandra*

Family: BORAGINACEAE
Common name: **Borage**
Number of genera: 146
Number of species: 2000
Origin: Worldwide
Plants: Trees, shrubs, and herbs

The fruits are drupes, nutlets, or capsules. Collect the fruits once they are ripe or dry. The capsules often require threshing and have fine silicate hairs that can cause skin irritation. Some species of seed have barbed hooks that attach to passing animals or objects.

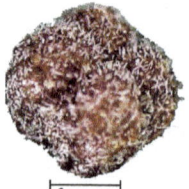

Bushy Heliotrope
Heliotropium tenuifolium

Camel Bush
Trichodesma zeylanicum

Family: CAESALPINIACEAE
Common name: **Bean**
Number of genera: 150
Number of species: 2200
Origin: Worldwide
Plants: Trees and shrubs with some herbs and climbers

The bean is formally a part of the family FABACEAE. The fruit is a legume (pod) and generally resembles the typical bean shape. They can be woody or papery thin, with one to numerous seeds, and can be spherical to flat and oval as well as cylindrical. The seed is reddish, grey, brown, or black, sometimes mottled.

Collect the whole pod individually or as a cluster, depending on the species. Once fully dry, they often require threshing to release the seed. Some species require very violent treatment to smash the pods. Wetting the pods and allowing time to dry before wetting again often helps open them.

Poinciana *Delonix regia*

Cassod Tree
*Cassia
siamia*

Bauhinia
*Lysiphyllum
hookeri*

Wild Tamarind
Lysiloma catisiliqe

Peltophorum
*Peltophorum
pterocarpum*

Leopard Tree
*Caesalpinia
ferrea*

**Cocky's
Tongue**
*Templetonia
retusa*

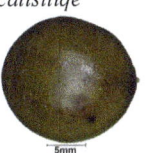

**Yellow
Nickers**
*Caesalpinia
bonduc*

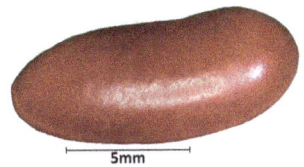

Batwing Coral Tree
Erythrina vespertillio

Family: CAPPARACEAE
Common name: **Caper**
Number of genera: 45
Number of species: 800
Origin: Tropical to subtropical
Plants: Trees and shrub with some climbers.

The fruit is an oval or cylindrical dry capsule that houses one to numerous seeds, or a fleshy berry containing one to numerous seeds in a sticky cream-colored pulp.

The seeds are round and flattened and can be smooth or pitted. Collect the fruits individually once fully ripe. Wash away the sticky pulp from the seed and wipe dry with a cloth before drying and storing. The seed will last many years if left within an undamaged dry berry.

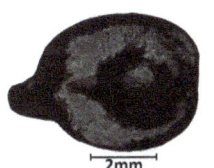

Silver Cassia
*Senna
artimisiodes*

Tagasaste
*Cystus
prolifer*

Split Jack
*Capparis
lasiantha*

Wild Orange
*Capparis
mitchelii*

Parkinsonia
Parkinsonia aculeate

13

Family: CASUARINACEAE
Common name: **Sheoak**
Number of genera: 4
Number of species: 90
Origin: Asia Pacific region
Plants: Trees and shrubs

The fruit is a hard, woody cone. The seeds are winged, ranging in size from 3 - 17mm, and are light grey / brown to black in color.

Collect the cones whole and allow time for drying and release of the seed. The seed can be cleaned by removing the empty cones. Any additional chaff can be removed by the use of a sieve.

River Sheoak
Casuarina cunninghamiana

Black Sheoak	**Swamp Sheoak**
Allocasuarina litoralis	*Casuarina obesa*

Family: CELASTRACEAE
Common name: **Celastra**
Number of genera: 50
Number of species: 800
Origin: Tropical to subtropical
Plants: Trees and shrubs

The fruit is a capsule, drupe, or berry. They may or may not open upon maturity. Capsules can be oval to cylindrical or three sided. The seeds are oval-shaped and covered by a colorful, sticky flesh which is yellow, orange, red, or brown in color. The

seed can also be flat, winged, and papery.

Collect the whole fruit once fully ripe. Gather the berries and drupes and clean them of their flesh and pulp before leaving them to dry.

The seeds of capsules can be collected by scraping the seeds out of the capsules and allowing time to dry.

Bullock Bush
Denhamia oleaster

Family: COCHLOSPERMACEAE
Common name: **Kapok**
Number of genera: 2
Number of species: 25
Origin: Tropical
Plants: Trees and shrubs

The fruit is a capsule that usually splits lengthwise upon drying. The seeds are kidney-shaped and hard, usually colored dark brown to black. Collect the whole capsule as it dries and thresh before sieving the seed. The fibrous hairs may cause irritation to sensitive skin.

Kapok Bush
Cochlospermum gillivraei

Family: COMBRETACEAE
Common name: **Indian Almond**
Number of genera: 20
Number of species: 500
Origin: Tropics and subtropics
Plants: Trees and shrubs

The fruit is a winged or ridged drupe containing one seed / kernel. The whole fruit is collected individually. No cleaning or threshing is required other than removing any dirt and debris.

Beach Damson
Terminalia muelleri

Family: CUCURBITACEAE
Common name: **Cucurbit**
Number of genera: 120
Number of species: 830
Origin: Tropical to subtropical
Plants: Herbaceous scrambling vines

The fruits are classified as pepos. They are very variable: small to large, fleshy or dry. They can house the seed for extended periods. There can be one to numerous seeds. These seeds can be small to large and are often flattened and sculptured; they can also be smooth or winged.

Collect the whole fruit once fully ripe. The gourds can be collected when the vine is dead and the fruits fully dried and hard. Some immature seed is usually found within the ripe fruit—these should be discarded.

The seed often sheds a thin papery layer once fully dry. Clean away this chaff before storage.

Balsam Apple
Momordica balsamina

Family: CUPRESSACEAE
Common name: **Cypress**
Number of genera: 19
Number of species: 120
Origin: Temperate regions of the northern Hemisphere
Plants: Trees and shrubs

These plants bear male and female cones. The fruiting cones split along each section, spiraling around the cone from top to bottom. The seeds are flattened and hard, usually with one or two wings. They are shades of reddish-brown. Collect the whole cone as it begins to dry and allow time for the entire cone to release its seeds before beginning the sieving process.

Cypress Pine
Callitris intratropica

Family: CYCLANTHACEAE
Common name: **Cyclanthus**
Number of genera: 12
Number of species: 180
Origin: Tropical to subtropical
Plants: Trees, shrubs and climbers

The fruits are berries or clusters of berries. The seeds sometimes have a winged edge. Collect the whole fruits individually and gently squeeze out the pulp into a sieve and lightly press the pulp through the sieve with your fingertip and slow running water. Allow time to dry before sieving out any remaining chaff.

Panama Hat Plant
Carludovica palmate

15

Family: CYCADACEAE
Common name: **Cycads**
Number of genera: 1
Number of species: 20
Origin: Southern Hemisphere
Plants: Trees

These plants are protected in the wild and special permits must be obtained before any collections can be undertaken.

The seeds are large and oval: 30mm x 25mm.

Collect the seeds individually. No cleaning or threshing is required other than removing any dirt and debris.

Zamia Palm
Cycas media

Prince Sago
Cycas tiwaniana

Family: DIPTEROCARPACEAE
Common name: **Dipterocarpus**
Number of genera: 16
Number of species: 580
Origin: Tropics
Plants: Trees

The fruits are hard, woody berries, capsules, or nuts. Collect the fruits individually once ripe. The berries and capsules are difficult to open and require a hammer or similar tool to smash apart. Remove the seed, and clean away any chaff before drying and storing the seed.

Elephant Apple
Dellenia indica

Family: ELAEOCARPACEAE
Common name: **Elaeocarpus**
Number of genera: 12
Number of species: 605
Origin: Tropical to temperate
Plants: Trees and shrubs

The fruits can be fleshy or non-fleshy berries or drupes; when a capsule, they may open or not as they dry.

Collect the whole fruit and treat it depending on the fruit type. The drupes shown below had a thin layer of flesh which was removed.

Silver Quandong
Elaeocarpus augustifolius

Blue Quandong
Elaeocarpus grandis

Family: EUPHORBIACEAE
Common name: **Spurge**
Number of genera: 300
Number of species: 7500
Origin: Worldwide
Plants: Shrubs and herbs with a white, milky sap

The fruit is usually a dry capsule that splits lengthwise upon drying; it can also be a drupe or berry.

The seeds can be oval, triangular, or round and can be smooth or pitted. The drupes or berries are collected once ripe, and the capsules are collected as they split.

White Currant Bush
Flueggea virosa

Quinine
Petalostigma pubescens

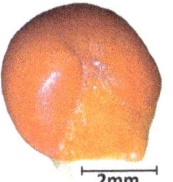

Buttonwood
Glochidion sumatranum

Cheese Tree
Glochidion ferdinandi

Prostrate Spurge
Chamaesyce sp.

Bellyache Bush
Jatropha gossypiifolia

Family: GUTTIFERAE, CLUSIACEAE
Common name: **St John's Wort**
Number of genera: 37
Number of species: 1610
Origin: Worldwide
Plants: Trees, shrubs, herbs, and climbers

The fruits can be fleshy or dry drupes, berries, or capsules. The seed can be held within the fruit or released upon maturity. The seed is sometimes winged. Collect the fruit and clean, depending on the fruit type encountered.

Beauty Leaf
Calophyllum inophyllum

Family: HERNANDIACEAE
Common name: **Hernandia**
Number of genera: 3
Number of species: 54
Origin: Tropical to subtropical
Plants: Trees, shrubs and climbers

The fruit is a samara that is non-fleshy. Collect the samaras individually or in bunches. The wings can be threshed off, leaving only the seed. Sieve out any chaff before storing the seed.

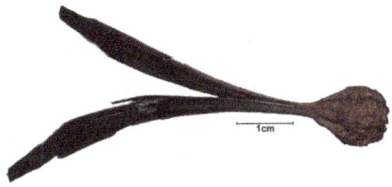

Stinkwood
Gyrocarpus americanus

17

Family: LAURACEAE
Common name: **Laurel**
Number of genera: 50
Number of species: 2000
Origin: Tropics to subtropics
Plants: Evergreen trees and shrubs with some climbers

The fruit is usually a berry or drupe. The seed is often enclosed in a sticky, succulent fruit. It can be oval and partly flattened. The seed is creamy white to brownish-black. Collect the fruits individually, and remove the flesh and pulp. The seed is short-lived and should be planted soon after collection.

Dodder Laurel
Cassytha filiformis

Family: LEEACEAE
Common name: **Leea**
Number of genera: 1
Number of species: 70
Origin: Tropics
Plants: Trees, shrubs and herbs

The fruit are fleshy berries. Collect the whole ripe berries and gently squeeze the pulp into a sieve before lightly pressing the pulp through the sieve with your fingertip and slow running water. Dry the seed thoroughly before storage.

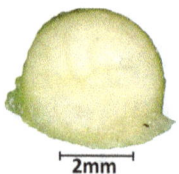

Anka Dora Holly
Leea indica

Family: LYTHRACEAE
Common name: **Myrtle**
Number of genera: 30
Number of species: 600
Origin: Worldwide
Plants: Trees, shrubs and herbs

The fruits are non-fleshy capsules that may or may not open upon drying. The seed is either winged or wingless. Collect whole capsules as they begin to dry and thresh lightly before sieving out the seed.

Crepe Myrtle
Lagerstromia archerana

Family: MAGNOLIACEAE
Common name: **Magnolia**
Number of genera: 7
Number of species: 225
Origin: Tropical to subtropical
Plants: Evergreen trees and shrubs

The fruits are cone-like capsules with a few seeds that are orange to red in color. Collect whole capsules and allow time for them to dry and release their seeds before sieving out any chaff.

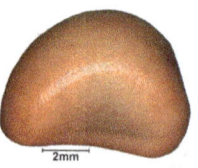

Golden Champaca
Michelia champaca

Family: MALVACEAE
Common name: **Mallow**
Number of genera: 75
Number of species: 1500
Origin: Worldwide
Plants: Shrubs and herbs

The fruits are capsules which split from the top downwards. They are sometimes berries. The capsules can be harvested once dry. The seeds are somewhat kidney-shaped, and occasionally winged or with horns. They are brown or black in color.

Collect the capsules using gloves and thresh before sieving out the seed. The chaff can cause extreme skin irritation and may affect people with breathing difficulties.

The fruit is a capsule, drupe or berry, or sometimes a nut. The seed can be winged or wingless, round or oval. Collect the fruits individually.

The capsules split as they dry, releasing their seed. These should be gathered as they begin to split and placed in a bag to dry. Remove the outer casing of the capsule and store the clean seed.

The berries and drupes are collected when fully ripe. Remove the seeds from the pulp and give them time to dry before sieving out any chaff.

The nuts are collected individually and stored. No cleaning is usually required.

Rose of Sharon
*Hibiscus
syriacus*

Pima Cotton
*Gossypium
barbadense*

African Mahogany
Khaya senegalensis

Portia Tree
*Thespesia
populnea*

**Sturt's
Desert Pea**
*Alyogine
hakeifolia*

Emu Apple *Owenia acidula*

Family: MELIACEAE
Common name: **Mahogany**
Number of genera: 51
Number of species: 550
Origin: Tropical and subtropical
Plants: Trees and shrubs with some herbs

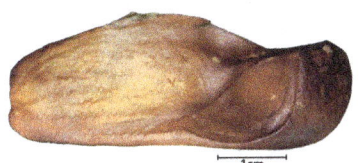

Spanish Mahogany
Swietenia mahoganii

19

Monkey-puzzle Nut
Xylocarpus granatum

Indian Mahogany
Toona ciliate

Family: MENISPERMACEAE
Common name: **Moonseed**
Number of genera: 75
Number of species: 520
Origin: Tropical and subtropical
Plants: Woody vines with some trees shrubs and herbs

The fruits are drupes or an aggregate of drupes. The seed is usually round or oval. Collect the fruit individually or as clusters once fully ripe. The flesh is usually removed from the seed easily. Wipe away any remaining pulp from the seed with a clean cloth. Dry the seed on a sheet that will not sweat or cause moisture to form.

Laurel-leafed Hypserpa
Hypserpa laurina

Family: MIMOSACEAE
Common name: **Mimosa**
Number of genera: 60
Number of species: 3000
Origin: Tropical to temperate
Plants: Trees and shrubs with some herbs

This family is formally incorporated within the family FABACEAE. The fruits are classified as "legumes," which are bean pods. They can be flat or cylindrical in shape and can be hard and woody through to thin and papery. The pods may have one to numerous seeds. The seeds may be spherical to flat and oval and are found in many colors, ranging from white, yellow, orange, red, grey, and brown to black, including mottled.

The genus *Acacia* do not have true leaves, but rather enlarged leaf stalks called phyllodes.

Collect the whole pods individually or in bunches and allow time to dry fully. Some species require wetting several times before they will release their seeds. Others open explosively and spread their seeds over several meters—a covering of material or shade cloth is required to contain the seeds.

Once dry, threshing is usually required to free the seed fully before sieving out the chaff.

Halls Creek Wattle
Acacia cowleana

Coobah   **Hickory Wattle**
Acacia salicina   *Acacia falcate*

Siris Tree
Albizia lebbeck

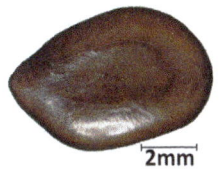

Leucaena
Leucaena leucocephala

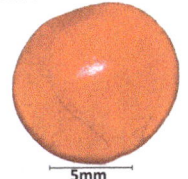

Basket Flower
Adenanthera pavonina

Matchbox Bean
Entada rheedii

Family: MORACEAE
Common name: **Fig**
Number of genera: 53
Number of species: 1400
Origin: Tropical to subtropical
Plants: Trees, shrubs, herbs and some climbers

The fruits are syconiums (fig), soro-sises (mulberry), berries, or drupes. The seeds are variable in size and shape, ranging from very small (figs and mulberry) to large (*Artocarpus* sp.). The seed can also be round, curved, or straight.

Due to the variable fruits, the collection of the seed will vary, depend-ing on what fruit is collected. The specimen sought may be a single dry seed or numerous seed enclosed in pulp.

Be sure to identify what type of fruit are being collected (Introduction), and base the cleaning and/or extraction of the seed on what you encounter.

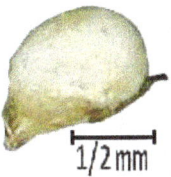

Long Leafed Fig
Ficus longifolia

White Fig
Ficus virens

Family: MYOPORACEAE
Common name: **Emu Bush**
Number of genera: 5
Number of species: 200
Origin: Southern Hemisphere, mostly mainland Australia
Plants: Small shrubs

The fruit is a drupe, and is oval to elliptical, ranging in size from 2 to 15mm. The fruit has a characteristic thin tail at the base that is easily bumped off. Collect the fruit individually from the bush where possible: as the fruit falls quickly once ripe, they will often be scattered all over the ground. The fruits or seeds collected from the ground should be thoroughly cleaned before being stored.

Creek Wilga
Eremophila bignoniiflora

21

Family: MYRSINACEAE
Common name: **Myrsine**
Number of genera: 35
Number of species: 1000
Origin: Tropical to temperate
Plants: Trees, shrubs, and climbers

The fruits are drupes and berries containing one to many seeds. Collect the fruits individually and clean the larger seeds by hand under running water. Collect the small seed using a fine sieve, lightly pressing the pulp through the sieve with your fingertip and slow running water.

Native Coralberry
Ardisia pachyrhachis

Family: MYRISTICACEAE
Common name: **Nutmeg**
Number of genera: 19
Number of species: 300
Origin: Tropical
Plants: Small trees and shrubs

The fruits can be fleshy to non-fleshy drupes or berries. The seeds are collected when the fruit is fully ripe. The non-fleshy drupes, like the following nutmeg, are easily collected, with no cleaning required. The fleshy drupes and berries should have any pulp removed before cleaning and storing the seed.

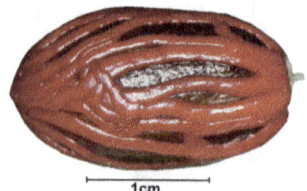

Queensland Nutmeg
Myristica insipida

Family: MYRTACEAE
Common name: **Myrtle**
Number of genera: 155
Number of species: 3500
Origin: Australasia
Plants: Trees and shrubs

All of the myrtles have oil glands in their leaves and are predominantly evergreens. The fruits are drupes or capsules that can be papery or woody. The capsules split at the front upon maturity, releasing the seed. Some capsules may remain unopened for several years. The seed can be saucer-shaped, wingless, have a wing all around or at one end, be pyramid, crescent-shaped, or straight. The seed color ranges from reddish-brown to brown or black.

The capsules are collected after they turn brown or the lip of the capsule changes color. Collect the fruits once mature and spread on a ground sheet until the seeds are released, then sieve the seed out. Most species of these seeds are long-lived (twenty years or more), while others only remain viable for 1 to 2 years.

The drupes have a limited life span of a few weeks, but can be stored in a refrigerator for a month or so.

Pink Marri
Eucalyptus calophylla

River Red Gum
Eucalyptus camaldulensis

Liniment Tree
Asteromyrtus symphyocarpa

Swamp Box
Lophostemon grandiflorus

Fine Tea Tree
*Agonis
parviceps*

Bottlebrush
*Callistemon
hinchinbrook*

**Silver
Cadjeput**
*Melaluca
argentia*

Silky Tea-Tree
*Leptospermum
myrsinoides*

Swamp Bloodwood
Corymbia ptychocarpa

**Golden
Penda**
*Xanthostemon
chrysanthus*

**Daintree
Satinash**
*Syzygium
monospermum*

Family: OLEACEAE
Common name: **Olive**
Number of genera: 30
Number of species: 600
Origin: Worldwide
Plants: Trees, shrubs, and woody vines

The fruit is a capsule, drupe, berry or winged nut, and usually has only one seed. Collect the fruit and remove any flesh before cleaning with a dry cloth and allowing time to dry. Seeds such as the following jasmine can stain clothing, so handle with care or wear a protective apron and gloves. Try using it as a dye.

Native Jasmine
Jasminum didium

Family: ONAGRACEAE,
OENOTHERACEAE
Common name: **Primrose**
Number of genera: 17
Number of species: 675
Origin: Subtropical to temperate
Plants: Shrubs and herbs

The fruits are capsules, berries, and nuts. Seeds are oval, slightly ridged lengthwise, and often numerous, with some bearing small tufts of hair. Many escape cultivation to become minor weeds. Collect the capsules and thresh once fully ripe. The berries can be lightly pressed through a sieve using your fingertip and slow running water.

Fire Weed *Epilobium* sp.

Family: PANDANACEAE
Common name: Pandanus
Number of genera: **3**
Number of species: **900**
Origin: Tropics
Plants: Trees, shrubs, or climbers

The fruit is a large, globular, ellipsoid, or cylindrical aggregate of drupes with large cylindrical seeds.

Collect the whole fruit once mature or the individual segments as they fall from the plant. The seeds cannot be removed from each segment, so they are stored whole. The segments germinate reasonably well when planted fresh.

Pea Bush
Sesbania juponica

Red Moneywort
Alysicarpus rugosus

Screw Pine
Pandanus tectorius

Butterfly Bush
Petalostylis labicheoides

Indian Beech
Pongamia pinnata

Family: PAPILIONACEAE
Common name: **Legume**, **pea,** or **pulse**
Number of genera: 400
Number of species: 12000
Origin: Worldwide
Plants: Trees, shrubs, herbs, and vines

The legumes are formally known as FABOIDEAE, FABACEAE, or LEGUMINOSAE.

The fruit is a pod (legume) with two equal halves that opens once dry, sometimes dispersing the seed with intense force.

The seed size and shape is variable; generally, it is a cylindrical or flattened oval design, which sometimes has hairs. Collect the pods and thresh if required before sieving out the seed.

Sometimes a covering is required to contain seed from explosive pods. Moisten hard woody pods to help release the seeds.

Rosewood *Tipuana tipu*

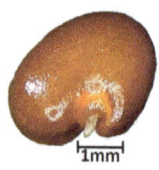

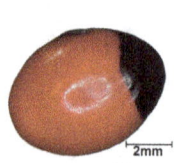

Rattlepod
Crotalaria dissitiflora

Crab Eye Vine
Abrus precatramia

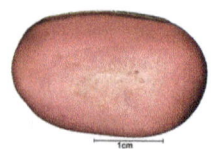

Sword Bean
Canavilia gladiata

Silverbush
Sophora tomentosa

Family: PASSIFLORACEAE
Common name: **Passionfruit**
Number of genera: 10
Number of species: 500
Origin: Tropical to subtropical
Plants: All are vines

The fruits are thin-walled capsules or berries. The seeds are spearhead-shaped and pitted, and are usually surrounded in a fleshy pulp. The seed is brown to black in color.

Collect the whole fruit and cut open to reveal the seeds. Scoop the seeds and pulp into a sieve and press the pulp lightly through the sieve with your fingertip and slow running water. Dry thoroughly before storage.

Wild Passionfruit
Passiflora foetida

Family: PINACEAE
Common name: **Pine**
Number of genera: 11
Number of species: 250
Origin: Temperate Northern Hemisphere
Plants: Trees and shrubs

The fruit is a cone-containing winged or partially winged seeds. Collect the cones as they begin to split and place on a ground sheet to dry. The segments will open one or a few at a time until all the cone segments are fully open. Sieve out the seeds as required, discarding unripe, small, or misshapen seeds.

Radiata Pine *Pinus radiata*

Family: PIPERACEAE
Common name: **Pepper**
Number of genera: 8
Number of species: 3600
Origin: Tropical to subtropical
Plants: Small trees, shrubs, or scrambling vines

The fruits are fleshy, single-seeded drupes, usually growing in bunches. The fruit is collected and dried. The seed is not usually removed from the fruit; however, it can be removed when either fresh or dried if desired.

Giant Pepper Vine
Piper novae-hollandiae

Family: PITTOSPORACEAE
Common name: **Pittosporum**
Number of genera: 9
Number of species: 200
Origin: Tropical to subtropical of the Southern Hemisphere
Plants: Evergreen trees and shrubs

The fruit is a capsule or berry that splits from bottom to top, revealing a red pulp containing seeds that are variable in size and shape.

Collect the whole fruit as it begins to open. Remove the pulp and allow time for it to dry fully. The tacky flesh is not usually removed from the seed. Once the seed is dry, it can be lightly threshed and sieved of any loose material.

Weeping Pittosporum
Pittosporum Phyllyraeoides

Rusty Pittosporum
Pittosporum ferrugineum

Family: PROTEACEAE
Common name: **Protea**
Number of genera: 79
Number of species: 1700
Origin: Southern Hemisphere
Plants: Trees and shrubs

The fruits are woody structures that open into two equal parts. This sometimes occurs only after fire. The seeds are flat and papery, and may be with or without a wing. Seeds can be tan, brown, or black. Collect the woody capsules and place on a ground sheet to dry. Heat treatment is sometimes required to open seed pods. Sieve out any fine chaff and remove the empty capsules by hand. If the capsule is collected too soon, the seeds' viability can vary depending on its maturity.

Wickhams Grevillea
Grevillea wickhamii

Silver Oak
Grevillea decurrens

Grevillea decora

Family: RHAMNACEAE
Common name: **Buckthorn**
Number of genera: 55
Number of species: 900
Origin: Worldwide
Plants: Trees and shrubs

The fruit is a drupe or capsule or sometimes a nut or samara (*Ventilago*). Each fruit has one seed. It is usually brown to black, round and pitted.

The seed are collected individually or in bunches. The seed of the White Ash shown below is covered with a thin orange layer which is removed upon cleaning.

White Ash *Alphitonia petrii*

The flesh of drupes is removed from the seed before drying and storage.

Chinee Apple
Ziziphus mauritiana

This samara fruit is collected from the tree in bunches. The wing can be removed by threshing leaving only the seed for storage.

Supple Jack
Ventilago viminalis

Family: RHIZOPHORACEAE
Common name: **Mangrove**
Number of genera: 15
Number of species: 120
Origin: Tropical to subtropical
Plants: Trees and shrubs

The fruits are non-fleshy berries, drupes, and occasionally capsules. Many of the fruits are short-lived and require tidal mud flats to germinate.

Mangrove
Ceriops decandra

Red Mangrove
Rhizophora stylosa

Family: ROSACEAE
Common name: **Rose**
Number of genera: 122
Number of species: 3350
Origin: Worldwide
Plants: Trees and shrubs with some herbs.

The fruits of roses are extremely varied and can be pomes, drupes, hips, or other forms of fruit. The seeds are also variable in size and shape, ranging from a single large seed to minute dust-like seeds.

Collect the mature fruits and remove the seeds depending on the fruit type.

Rose-leaf Bramble
Rubus queenslandicus

Family: RUBIACEAE
Common name: **Madder**
Number of genera: 650
Number of species: 5000
Origin: Tropical to subtropical
Plants: Trees, shrubs, herbs, and vines

The fruits can be dry capsules or fleshy berries or drupes. They contain two to numerous seeds which are often embedded in a succulent mass. The seeds can be highly variable in size, shape, and coloration.

Collect the capsules as they begin to dry and split. The seeds fall freely from the capsules and little, if any, sieving is required.

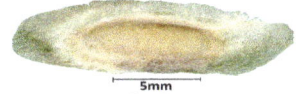

Teak *Flindersia* sp.

Remove the flesh and pulp from the drupes and wipe the seeds clean before drying and storing

Brown Gardenia
Atractocarpus fitzalaii

The small seeds from berries can be cleaned using a sieve by very gently pressing the pulp through a sieve using a spatula and slow running water.

Leichhardt Tree
Nauclea orientalis

Family: RUTACEAE
Common name: **Rue**
Number of genera: 150
Number of species: 1500
Origin: Tropical to temperate regions
Plants: Evergreen trees or shrubs

The fruits are drupes, berries, or non-fleshy fruits that split into parts. The seeds are compressed, oval, kidney shaped, or elliptical. They are brown to blackish in color.

The non-fleshy fruits are collected and allowed to dry before threshing and sieving of the seeds.

The flesh from drupes and berries is removed and the seeds wiped clean with a dry cloth before drying and storing.

Pink Evodia
Melicope elleryana

Mock Orange
Murraya ovatifoliolata

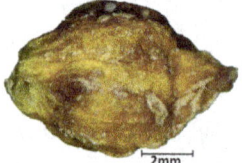

Wood Apple
Limonia acidissima

Family: SANTALACEAE
Common name: **Sandalwood**
Number of genera: 38
Number of species: 400
Origin: Worldwide
Plants: Trees, shrubs, and herbs

These plants are mainly root parasites and require a host for part or all of their lives. The fruit is a drupe or nut containing one seed. Collect the fruit individually and remove any flesh. Several species have sweet, edible flesh.

The seed stores well and requires scarification prior to planting. A temporary host plant, such as grass, must be provided soon after germination occurs or the seedling will die.

Broad-leaved Cherry
Exocarpos latifolius

Northern Sandalwood
Santalum lanceolatum

Family: SAPINDACEAE
Common name: **Soapberry**
Number of genera: 150
Number of species: 2000
Origin: Tropical and subtropical
Plants: Trees and shrubs

The fruits are variable and can be drupes or capsules. The *Dodonaea* is a three-sided capsule; its seed is round to oval or elliptical, and white to brown, red and black. The fruit of *Atalaya* is a samara with an oval light-brown seed. Many species of SAPINDACEAE produce fruit of economic importance and are also covered in the chapter "Fruits." The seeds are collected using a variety of methods, depending on the fruit type.

Capsules and samara are usually collected in bunches and threshed before removal of the chaff.

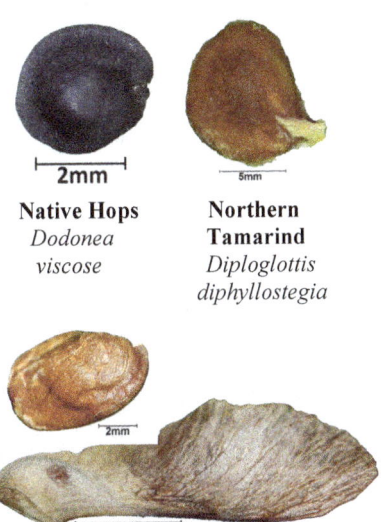

Native Hops
Dodonea viscose

Northern Tamarind
Diploglottis diphyllostegia

Whitewood
Atalaya hemiglauca

Family: SCROPHULARIACEAE
Common name: **Figwort**
Number of genera: 250
Number of species: 5000
Origin: Worldwide
Plants: Shrubs and herbs

The fruits are small capsules that split from the top or sides producing few to many small seeds. The seeds can be arrow- or kidney-shaped, rectangular, or oval, with or without a wing. All are pitted and are light brown or brown to black in color. Collect the whole capsule and allow time to open fully before threshing out the seed. Sieving of the chaff may be required, depending on the species.

Express Tree
Paulownia tomentosa

Family: SOLANACEAE
Common name: **Nightshade**
Number of genera: 90
Number of species: 2800
Origin: Americas
Plants: Small trees, shrubs, herbs, and vines

Many of the genera from this family are important agricultural plants. The fruits are fleshy capsules or berries which can be harvested upon maturity.

The seeds are flattened, curved, and oval, often with pitting; the color ranges from yellow, orange, grey, brown to black.

Collect the fruits once fully ripe. The capsules may or may not open upon maturity. Dry capsules can be threshed, whereas fleshy capsules can be cut open and the seeds removed manually.

The pulp of berries can be pressed through a sieve, or placed in a container of water and left for a week or so to ferment, before being sieved clean.

1/2 mm

Wild Tobacco
Nicotiana megalosiphon

29

Family: STERCULIACEAE
Common name: **Sterculia**
Number of genera: 65
Number of species: 1000
Origin: Tropical to subtropical
Plants: Trees and shrubs

The fruit is a berry or a follicle (opens on one side revealing the seed). The seeds are often covered with a smooth or hairy, thin membrane. The seed is yellow, brown, or black in color. The hairs on the seeds of some species can cause extreme irritation to the skin and throat, and should be collected with care. Collect the fruits and remove the seeds individually.

Flame Tree
*Brachychiton
acerifolius*

**Brown
Kurrajong**
*Commersonia
bartramia*

Looking Glass Tree
Heritiera littoralis

Family: STRELITZIACEAE
Common name: **Bird of Paradise**
Number of genera: 3
Number of species: 7
Origin: Tropical to subtropical
Plants: Large trees and shrubs

The fruit is a capsule with conspicuous hairy seeds. Collect the capsules individually and remove the seeds with care. Light threshing may be required to dislodge stubborn seeds. Sieve to remove any chaff.

Bird of Paradise
Strelitzia nicolai

Family: TILIACEAE
Common name: **Linden**
Number of genera: 50
Number of species: 450
Origin: Worldwide
Plants: Trees, shrubs, and herbs

The small fruits are capsules, drupes, or nuts containing one to several seeds. Collect all fruiting forms and allow time to dry before threshing. The capsules will release their seeds upon threshing. The flesh from the drupes should fall away. The nuts should not require threshing. Sieve out the seed and remove the chaff.

Dogs Nuts
Grewia sp.

Wild Jute
*Corchorus
trilocularis*

Family: URTICACEAE
Common name: **Nettle**
Number of genera: 56
Number of species: 2600
Origin: Worldwide
Plants: Trees, shrubs, herbs and climbers

The fruit is a single-seeded dry capsule or nut; it can also be a dry or fleshy drupe. Collect the dry capsule and thresh prior to sieving out the chaff. The nuts can be stored once fully dry. The drupes should have any flesh and pulp removed before being left to dry.

Nettle *Urtica dioica*

Family: VERBENACEAE
Common name: **Verbena**
Number of genera: 100
Number of species: 3000
Origin: Tropical and subtropical
Plants: Trees, shrubs, herbs and vines that are usually hairy

The fruit is an oval single-seeded drupe or dry berry which slips into 2 to 4 seeds upon drying. Collect the fruit once fully ripe. Remove any flesh from the drupes and lightly thresh. Allow time to dry before sieving out any chaff.

White Beech **Chaste Tree**
Gmelina *Vitex* sp.
dalrymplena

Family: VITACEAE
Common name: **Grape**
Number of genera: 15
Number of species: 700
Origin: Tropical to subtropical
Plants: Small shrubs or woody vines

The fruit is a fleshy berry often containing four seeds. The pulp can be squeezed onto a sieve and then lightly pressed through the sieve to leave the seed. Place the seed on a clean surface to dry, preferably not plastic, as this tends to sweat and cause mold.

Wild Grape
Ampelocissus acetosa

Family: XANTHORRHOEACEAE
Common name: **Grass Trees**
Number of genera: 10
Number of species: 100
Origin: Endemic to Australia, with the exception of two species in New Guinea and one in New Caledonia
Plants: All perennial grass-like plants

The flower spikes produce numerous fruit capsules which split along their lengths.
   The seeds can be round, oval, or flattened, and range in color from orange, brown, red to black.
   Collect the whole flower spike, or the individual capsules and spread on a ground sheet to dry before sieving out the seed.

Grass Tree *Xanthorrhoea* sp.

Family: ZAMIACEAE
Common name: **Zamia Palm**
Number of genera: 8
Number of species: 80
Origin: Africa, America, and Australia
Plants: Trees

These plants are protected in the wild and special permits must be obtained before any collections can be undertaken.

The fruit is a cone that produces numerous fleshy orange-colored seeds. Collect the individual seeds as they ripen along the cone.

Wild Sago *Zamia integrifolia*

Sago Palm *Zamia* sp.

Fruiting Cone of Zamia Palm

Fruiting Cone showing colorful seeds

CHAPTER 3
HOW TO THRESH
AND CLEAN SEED

Threshing

First, let's look at the term **threshing**.

Threshing is the term used for the breaking up and separation of the dried seed pods, seed heads, or capsules from the seed. The waste from the seed is called **chaff**. It is after threshing that the actual cleaning of seed occurs. These are two separate activities. Threshing of one form or another is usually required prior to cleaning.

With the use of some common household items, the threshing of seed is not very difficult. Sieves, colanders, tea strainers, buckets, and bowls can all be employed to get the job done—even by professional collectors. Additional equipment for enthusiastic collectors includes a mechanical thresher or a garden blower-vac.

Threshing can be undertaken by a number of means, including manual and mechanical methods. Both are effective, although an appropriate method should be sought depending on the amount of material that requires threshing. There is little point in using a mechanical thresher for a handful of material; conversely, you really don't want to thresh a truckload of material by hand!

Hand Threshing

As implied by the name, this is the use of your hands to crush the chaff from around the seed. This can be undertaken by placing the material in a bucket or bowl, and crunching it up until the seed is freed from the capsules. Gloves are generally required for this method, as the action required can hurt your hands.

Unfortunately, this is not always an effective technique for many of the harder woody species, as the pods are simply too difficult to crush manually.

After you have finished hand threshing the seed / chaff material, gently but firmly swirl the contents around in a bucket or bowl. The usually heavier seed will work its way to the bottom, allowing the bulk of the chaff to be removed and discarded. This makes the cleaning process easier. Should the seed and chaff not separate; cleaning will be more complicated and time-consuming.

Threshing with Sieves

Threshing can be undertaken by simply pressing the collected material through an appropriate sieve. This works well for many species with relatively brittle capsules or pods.

To determine which sieve is best suited for the material being threshed, select a sieve with holes slightly larger than the seed.

Good quality gloves are recommended when using this method, as injuries can occur from small sticks, thorns, etc. piercing the skin.

Attaching a sieve to a child's swing can aid greatly in this method of threshing, as it enables you to use a much larger sieve.

Manual, hand operated thresher

Threshing with a Towel or Rag

This method is restricted for the seed of fruit and berry species where the outer layer of the seed peels away like a skin. The cucurbits are one such group where this method is well-suited.

To use this method, rub the seeds between two layers of material, such as calico or rags. Place only small amounts of seed between the materials at a time, and rub back and forth to remove the chaff.

After you are satisfied the seed and chaff are separated, pour everything into a container to be cleaned.

Mechanical Threshing

There are many mechanical threshers on the market, both manual and motorized. Regardless of type, they are expensive to purchase. You would need to be doing large quantities of threshing to make the purchase of one worthwhile.

Cleaning Seed

The cleaning of seed after it has been threshed can be a very rewarding experience, as you get to see the end product of your labors. Cleaning, at its most basic, needs only to involve the use of two bowls and the wind; at the more complicated end of the spectrum, it may employ specialized equipment, such as sieves or a motorized cleaner.

Regardless of what equipment is required, cleaning the seed helps reduce the space required for storage, reduce or eliminate pest problems, and makes replanting much easier.

Two Bowl Method

The most simple and inexpensive way to clean seed is the aptly named Two Bowl Method. First, find two suitable bowls; these can be medium-sized plastic food containers, 10-liter water buckets, etc. Use your imagination!

Slowly tip the uncleaned seed (seed and chaff mixture) from the first bowl into the bowl underneath, allowing the wind to blow the chaff away. Wind can be provided either naturally or via a fan. The distance between the bowls will require adjustment, depending on wind strength. Swap bowls and continue tipping the seed from one bowl to the other until you are satisfied with the cleanliness of the seed. In some cases, the seed will not become 100% clean, and a final clean using tweezers is required. Amaranth and eucalyptus are good examples of seeds that are left only partly clean using this method.

Sieves and Sieve Sizes

Having several sieve sizes is always helpful. Purchasing analytical quality sieves (as pictured) can be prohibitively expensive.

Flour sieves

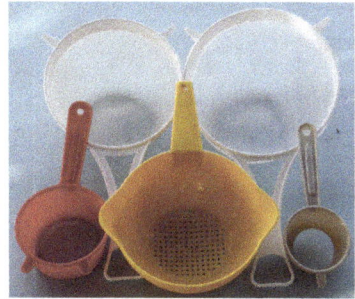

Assorted kitchen sieves

Analytical sieve

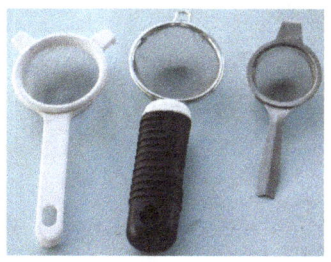

Assorted tea strainers

There are, however, alternatives to using expensive analytical sieves. The most cost-effective sieves can be found in almost every kitchen. These are basic flour and kitchen sieves, tea strainers, and colanders, all of which are available at your local store.

Other Sieves

There are many household items that can be used effectively as sieves for cleaning seed. Mosquito meshing comes in a handy size, whilst a colander is excellent for some of the larger seeds. Audio speakers, both car and home, provide mesh that makes an excellent sieve.

The material used for the mesh of a sieve is limited only by your imagination and creativity.

Speaker cover

Cleaning Pods and Dry Capsules

After capsules or pods have been threshed, the cleaning is greatly expedited using sieves. Often two or more sieves are used to separate the larger and smaller portions of the chaff from the seed.

Use several wide-mouthed bowls whilst sieving the seed, and have a bucket handy into which to place the waste. Doing this helps in chaff separation, and also contains any spillage in the event of an accident, making re-sieving easier.

Start with a sieve that has a mesh size several sizes larger than the seed. The reasoning behind this is to re-move the larger unwanted materials as efficiently as possible; this also reduces the volume of materials remaining to be cleaned.

After the bulk has been removed, use a sieve that is only slightly larger than the seed and slowly sieve the seed through. There will usually be several seeds remaining that are bigger than usual; collect these to add back to the cleaned seed once finished. Repeat this process until no large chaff can be effectively removed.

Now use a sieve slightly smaller than the seed and sieve out the smaller material. Often, small seed will be sieved out. Reject these, as they are usually immature or inferior. It is more beneficial to focus on the larger, healthy seed.

Repeat this process until the seed is as clean as possible. The final clean can be undertaken utilizing the wind or a mechanical cleaner.

Cleaning Fruits and Berries

The cleaning of fleshy fruits and berries is sometimes more interesting than it sounds, as many seed-bearing fruits have their own unique methods of seed extraction.

What can you do with seed that is covered in a sticky, tar-like substance? How do you remove the seed from a fruit that is excessively stringy and won't release the seed? The following information covers several of the common methods used in cleaning seed from fruits, berries, sticky pods, and other capsules that will inevitably be encountered.

Sticky Tar Substance

There are several ways of removing seed from pods and capsules when the seed is coated with a sticky, tar-like substance. Using dish detergent to clean the sticky substance works well. Make a solution twice as strong as you would for washing your dishes. Use disposable gloves, and rub the seeds together in the solution to dissolve and remove the stickiness. Several washes will often be required to end up with good quality, clean seed.

Mineral turpentine or kerosene will also remove the seed coating with some success; however, this presents the additional problem of removing a contaminating substance. This can be easily overcome by washing the seed in warm soapy water several times.

Remember, in both methods the seed must be thoroughly dried to avoid germination or fungal problems.

In most cases, the seed coating will dry naturally and can be cleaned at a later date. The problem with this is that it often takes about six months for the seed coating to dry, and insect pests can destroy all of the seed during this time.

Berries

Berries are fruits with more than one seed, such as grapes, gooseberries, and tomatoes. Some berries only have a few seeds, while others will contain hundreds. Regardless of how many seeds a berry has, they can usually be cleaned using a sieve and water.

If the size of the seed is unknown, you will need to gently cut open the berry and check the seed. To determine what size sieve will be required, select one that is slightly smaller than the seed. This will allow the pulp of the berry to be washed through the mesh. You may need to press the pulp through the sieve with your fingers or a small spatula.

Once the sieve is selected, place the entire berry or sections of it on the sieve. This process will depend on the size of the berry and sieve. Do not overload the sieve, as it only makes the task more difficult, and may result in the loss of seed, the failure of the sieve, or both.

With the water running slowly over the berry, gently press the pulp through the sieve. The seed should remain in the sieve. If the seed also passes through the sieve, select a smaller mesh size and start again.

Once the seed is considered clean, and the pulp is removed, place the seed on a clean, non-plastic surface and allow it to dry. This process can be expedited by dabbing the seed with a dry cloth, which can be a little tricky if the seed is small, as it tends to stick to the cloth and become a problem.

Spread the seed thinly and avoid mounds, as these are more likely to become moldy. As the seed dries, stir or mix at least daily to aid in thorough drying. This method works well for both small and large seeded species.

Many of the palm species are examples of easily-cleaned, dry-skinned drupes. They can be left a few days to dry and the skin will often peel off freely, leaving a clean seed.

Most stone fruit can be cleaned without much fuss, and any stringy flesh can be removed by cutting with a knife or scissors. The very small drupes with dry flesh can be cleaned whole. The entire fruit is left to dry out thoroughly prior to threshing the skins off and then cleaning with a cleaner or sieves.

Drupes

Drupes are single-seeded fruits such as plums, nectarines, and avocados. These can be either fleshy or not; this characteristic will determine how the seed is cleaned.

A quite pleasurable way to clean some of the fleshy edible species is to eat the flesh. Who can say no to avocado on toast?

The fleshy drupes can be cleaned with a sieve in the same manner as berries. However, cleaning the drupes without flesh can sometimes be much more difficult.

CHAPTER 4
THE STORAGE OF SEED

Introduction

Storing seed correctly is the most important step when considering how to collect, clean, and store seed. If seeds are stored correctly, they can last for years. However, if stored poorly, they will lose their viability and fail to germinate once planted.

An important point to remember here is that there are a number of seed-producing species that will only remain viable for a week, sometimes less. These seeds are mostly from fleshy tropical fruits, where there is plenty of moisture and germination occurs quickly. There are also many other plants in arid zones that produce dry capsules with seed that won't store for long. For these plants, storage is not an option, so the collection and cleaning of their seed should be undertaken with the intent of immediate propagation only.

There has been a great deal of research conducted over the last two centuries into the viability of seed under storage conditions, and many good research papers have been published on the findings. This research has shown that many of the seed species that are collected and cleaned correctly can be stored for use at a later date. This is good news for gardening enthusiasts, as it means that most of the plants they are collecting seed from can be preserved.

Regardless of the challenges involved in the storage of seed, it can be a very rewarding experience. All keen gardening enthusiasts should attempt to store the seed from their respective interests to plant in later seasons or exchange with other like-minded enthusiasts.

Take notes on when the seed was stored, how it was stored (in a box, in the refrigerator, etc.), and most importantly, when it was planted and if it germinated. This information on the viability of plant species in your district is very important to plant researchers, as it contributes to the general knowledge on plants worldwide.

Putting Seed into Storage

Only store the best quality seed; i.e., seed that is undamaged, free of defects, pests, chaff, and debris. There are occasions when you cannot collect the best specimens, as the seed simply isn't available due to weather conditions, seasonal factors, or just bad timing. When this occurs, all you can do is the best you can with what you have. Sometimes the seed will be suitable to propagate next season, and sometimes it will fail. This is all part of the enjoyment of collecting seeds.

Seeds should be stored once they are cleaned of all chaff, dirt, and fleshy dried pieces. This helps reduce the volume being stored and reduces the amount of contaminating materials in the seed. Contamination includes both pests and loose materials. Pests include insects, molds, and fungi, all of which can easily destroy stored seed.

Seed should not be put into storage on days when the humidity is high, as the seed may draw moisture from the atmosphere, which can lead to spoilage. This most commonly occurs when seed is stored in plastic bags.

Ideally, seed should be stored once fully dry. This sounds simple enough, but looks can be deceptive, and many seeds have been ruined due to eagerness. Most seeds take about two weeks to fully dry if maintained at an even, warm temperature. They may look dry on the outside, but are often still moist on the inside, so be patient. Many seeds reduce in size as they dry, so also look for this as it occurs.

Once the seeds are ready to be stored, take the time to examine them carefully for any problems, and remove anything that looks suspicious. "When in doubt, throw it out" is a simple expression that is well worth remembering.

Overcoming Seed Pest Problems

The storage of seed has always been an issue due to the many and varied circumstances that can arise without warning. Issues such as humidity, hot summers, cold nights, insects, molds and fungi, and many others add to the difficulty in storing seed successfully.

Seed for storage must be dry, free of dirt, and most importantly, free of insects and their eggs. The weevil eggs pictured below have hatched and the larvae have eaten this bean seed.

Weevils and Other Beetles

An excellent way of removing insect eggs from seed, especially large hard seed such as beans, is to make up a dilute solution of bleach (10ml/L) and soak the seed for ½ an hour. Re-

move and rinse the seed before thoroughly drying.

Fungicides and insecticides are readily available and can be used to treat seed where problems are too intense to overcome without their use. Take every possible precaution when using these products, as they are **dangerous**. Read the instructions on the label before using any product.

Weevils are one of several insect pests that eat seed. The photo above shows the damage that can happen if seed is stored poorly. A mature weevil is shown below.

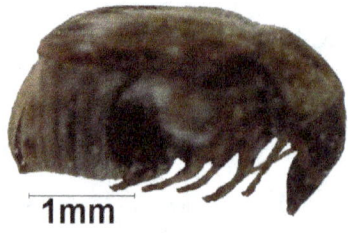

1mm

Weevils and other seed-boring beetles can range in size from almost microscopic to large, impressive specimens. Wood- and seed-boring beetles range in color from drab brown to highly colorful, and many are highly prized by collectors.

Moths

Moth larvae are a major economic pest in cereal crops throughout the world. A great deal of money is spent each year in trying to control the damage and losses caused by these small, troublesome insects.

The specimen shown below is about to pupate after consuming the bulk of a Bauhinia seed. Also shown is an immature larva that was inside the nearby seed when the pod was opened.

Being observant when cleaning your seed will prevent many of the problems that can arise with seed storage. However, it is always a good practice to check your seed regularly and remove anything that may look suspicious.

Fungi

Fungi are the group of plants that lack chlorophyll and leaves. They include molds, mushrooms, mildew, rusts, and smuts.

Of the pest problems encountered, problems with fungi are often the easiest to control and manage.

The problem with molds usually occurs if the seed is put into storage while it is still moist, or if the seed is stored under poor environmental conditions. Whatever the case, if the situation is discovered soon enough, the problem may be reversed, as it takes some time for molds and fungi

to penetrate the seed surface and cause damage.

If seed is found with mold growth, you should first remove all affected seed lots from the storage container, as you do not want to infect the remainder of your collection. The affected seed can either be disposed of or cleaned. If the seed is not too badly contaminated, the problem can be solved without disposal.

To clean mold from seed, gently rub the seed with a cloth moistened with bleach solution (10ml/L dilution) until the mold is removed. Staining often remains, but this should not affect the seeds' ability to germinate. This problem will have affected the seeds' long-term storage viability, so replacement should be considered as soon as possible.

Finally, fungicidal powders can be purchased that will inhibit fungal growth. These should be considered if this problem is likely to occur in your location. Remember to read the label and always follow the manufacturer's instructions—these products are **dangerous** when used incorrectly.

Temperature

Few people have access to expensive storage equipment that can maintain a stable temperature at the desired temperature setting. There are two issues here: the temperature at which the seed is stored, and the stability of this temperature. Seeds can be maintained at a low temperature of 2-4^{0}C for long periods, or they can be frozen for storage over centuries. It is the stability of the temperature that is of the greatest importance to the average collector.

As there is no way of maintaining a constant temperature without equipment, we will look at how to maintain a stable temperature using some commonly found items.

Changing temperatures are not good for seed storage. The constant fluctuations from the day and night cycle will cause the most rapid decline of seed viability. This is one problem that must be overcome if you intend storing seed for any length of time.

To achieve optimal results in saving seed, a suitable position with a stable temperature must first be selected. Avoid locations such as external walls and areas near amenities, air conditioners, fireplaces, or any other temperature-altering devices.

Once a storage location is selected, a suitable container in which to house the various lots of seed must be obtained.

Storage containers

Containers that are suitable include, but are not limited to: foam boxes, cardboard boxes, buckets, fiberglass containers, drums, etc. Some storage containers will require modifications to overcome temperature and humidity changes.

Foam boxes are the best choice, as they are already insulated and are often fully sealed. They can be found in numerous sizes, and usually have tight-fitting lids.

Cardboard boxes are good choices; however, finding thick boxes without gaps at the bottom and with tight-fitting lids is sometimes diffi-

cult. Additional cardboard can be glued in any places with gaps and can be used to thicken the base, top, and sides as required. Alternatively, line a cardboard box with polystyrene foam—thereby providing the best of both worlds.

Buckets and drums with tight-fitting lids can be used to store seed as long as the inside is properly insulated with foam or other suitable material. Remember that both the bottom and top require insulation. Otherwise, changes in temperature and humidity will still occur inside the container.

Polystyrene and cardboard are excellent insulation materials. Be extremely careful in choosing other insulating materials, such as roof and wall insulation. These items are not designed to be disturbed and are dangerous if handled incorrectly. If you require insulation materials for your chosen container that may be harmful, you should consult an industry expert.

Humidity

As changes in humidity can cause the loss of seed viability, the container that you have chosen to maintain a stable temperature must also be capable of preventing changes in humidity. This is one reason that the lid must be tight-fitting.

Humidity within any dwelling can change dramatically during the daily cycle, especially in tropical regions during the wet season. Even small changes in humidity can have unwanted results on the viability of your stored seed.

Silica gel is a desiccant, and is excellent for protecting your seed from moisture. Some gels are coated with cobalt chloride or other indicators that reveal when their usefulness has expired. You should check these regularly.

All silica gels can be dried and reused numerous times. To do this, place the gel packs on a tray and place in a cooling oven overnight. Ensure the oven is not too hot, as you do not want to burn the packaging of the gel packs.

Silica gels in one form or another can be purchased from chemists, hardware stores, nurseries, and supermarkets. There are a wide range of products on the market, and listing them all would be almost impossible. Water storage granules from nurseries are ideal.

Silica gel packs can also be collected from tablet bottles and within the packaging of electrical items and other miscellaneous products.

Remember that not all seeds can be stored, so choose your seeds carefully and avoid those that will present problems. As your experience grows, you can become more adventurous and attempt to store more difficult species, taking notes about the methods used and the outcomes.

Many of the rare species in collections around the world today have only survived because plant enthusiasts have saved them. Your efforts could lead to the saving of species for future generations to appreciate.

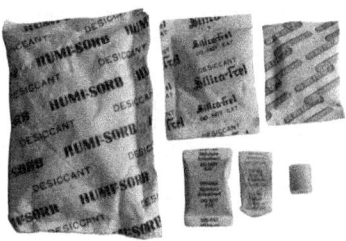

Ensure the gel packs are thoroughly dry before you use them.

Conclusion

The storage of seed can sometimes be challenging. However, the effort is very rewarding, and many varieties of seed can be stored successfully by anyone willing to take the time and effort.

Following simple suggestions, being patient, and taking care to avoid contamination and pests will all lead to the successful storage of seed.

REFERENCES

Bodkin F., *Encyclopaedia Botanica*. Cornstalk publishing, 1992

Brock J., *Top End Native Plants*. Dai Nippon Printing Co. Ltd, 1988

Cooper W., *Fruits of the Australian Tropical Rainforest*. Nokomis Edition Pty Ltd, 2004

Cronin L., *Australian Palms Ferns, Cycads and Pandans*. Envirobook, 2000

Flora of Australia. Volume 1. *Introduction* Australian Governemnt Publishing Service Canberra. 1981

Flora of Australia. Volume 3. *Hamamelidales to Casuarinales* Australian Governemnt Publishing Service Canberra. 1989

Flora of Australia. Volume 19. *Myrtaceae – Eucalyptus, Angophora* Australian Governemnt Publishing Service Canberra. 1988

Flora of Australia. Volume 29. *Solanaceae* Australian Governemnt Publishing Service Canberra. 1982

Jones D.L & Gray G., *Climbing Plants in Australia*. Reed Books Pty Ltd, 1988

Pilcher M, Davis L, Hurrion D., *Garden Terms*. Hamlyn, 1995

Wheeler J. R, Rye B. L, Koch B. L, Wilson A. J. G., *Flora of the Kimberley region*. Department of Conservation and Land Management,

Williamsn K.A.W., Native Plants, Queensland Volume 1. Printcraft, 1984

Williamsn K.A.W., Native Plants, Queensland Volume 2. Printcraft, 1988

Williamsn K.A.W., *Native Plants, Queensland Volume 3.* Printcraft, 1987

Williamsn K.A.W., *Native Plants, Queensland Volume 4.* CopyRight Publishing Co. Pty Ltd, 1999

Wikipedia, the free encyclopedia

Wrigley J.W & Fagg M., *Banksias, Waratahs & Gravilleas*. Collins Publishers Australia, 1989

Wrigley J.W & Fagg M., *Australian Native Plants*. Angus & Robertson Publishers, 1991

INDEX

www.ingramcontent.com/pod-product-compliance
Lightning Source LLC
Chambersburg PA
CBHW071137280526
45787CB00003B/1309